compiled by John Foster

Oxford University Press

also in this series:
A Very First Poetry Book
A First Poetry Book
A Second Poetry Book
A Third Poetry Book
A Fourth Poetry Book
Another First Poetry Book
Another Second Poetry Book
Another Third Poetry Book
Another Fourth Poetry Book
Another Fifth Poetry Book
A Scottish Poetry Book
A Second Scottish Poetry Book

Oxford University Press, Great Clarendon Street, Oxford OX2 6DP

Oxford New York
Athens Auckland Bangkok Bogotá Buenos Aires Calcutta
Cape Town Chennai Dar es Salaam Delhi Florence
Hong Kong Istanbul Karachi Kuala Lumpur Madrid
Melbourne Mexico City Mumbai Nairobi Paris São Paulo
Singapore Taipei Tokyo Toronto Warsaw

and associated companies in
Berlin Ibadan

ISBN 0 19 916053 8

First published 1985
Reprinted in hardback 1986, 1987, 1989, 1990, 1992
Reprinted in paperback 1986, 1989, 1992, 1994, 1996, 1998, 1999

Phototypeset by Tradespools Ltd, Frome, Somerset
Printed in Hong Kong

Contents

Books

Believe the golden stories in your head
And all the golden stories that you read,
For what you deeply feel
As true is much more real
Than all that can be counted, cold and dead.

John Kitching

And now

It's never now it's always wait until . . .
But this September Sunday, under Falkland Hill,
Cornstalks were stacked in rolls and rolls
And the sun struck gold in these ochre bales:
That moment hangs there still though I pass on
Chasing my shadow the length of the sun.

Alan Bold

Unsecret

Here's a thing you can't keep to yourself –
  The sunlight crashing on the sea
And smashed into a thousand smithereens
  That glitter crazily.

A black cloud moves across the sky, and now
  It snatches the hot sun from view
And sweeps the glittering fragments from the sea;
  But some light filters through;

And from each side of the black cloud white rays
  Of light reach out for all they're worth,
While, from the lower edge, black rays of rain
  Reach down to earth.

Edward Lowbury

The cave

I told the boatman to go deep,
Deep in the hollow cave
Where the black menace of
Each lifting wave
Licked with wet tongue and froth of lace
This barnacled and shadowed place.

I used to come here as a boy
And here at low tide felt the race
Of pulse
And throbbing in my throat,
Standing upon a shelf
Above the dark and sinuous moat
Of water,
Knowing that if my feet were caught
By mischance in a cleft of rock
I should be struggling in a lock
Of rising water.
The very thought
Of choking in that monstrous swell
Made me wheel round and yell
With panic.
I used to clamber, slide and slither back
Along that grey, forbidden track
And sob in safety on the tufts of thrift
Watching the cheated rollers shift
Slyly beneath my feet.

This day, in someone's boat
I catch again the terror of the place;
I clutch the collar of my coat,
And tell the boatman curtly, 'That's enough'.
The sweat is cold upon my face.

Gregory Harrison

Starfish

Went star-fishing last night.
Dipped my net in the inky lake
to catch a star for my collection.
All I did was splintered the moon.

Judith Nicholls

July evening

Wind over wet grass.
On the Hoe the lights of a fair
Trinket-bright against a drained sky.
Among the roundabouts, the easy rides,
A cylindrical cage spins;
Its speed is a gale
That pins men flat to its mesh.
Landward, the big wheel turns
High as rigging, the Drake stands
Sturdy on his marble plinth.
His bronze eyes look seaward, beyond
The twentieth century candyfloss,
The steady smokeless lamps.
To the real hungers of long voyaging,
The sleepless hours of storm, the swing
Of lanterns feeble in the dark,
The cry of strained timbers withstanding
The winds of uncharted seas.

Pamela Gillilan

Sunset

The sun spun like
a tossed coin.
It whirled on the azure sky,
it clattered into the horizon,
it clicked in the slot,
and neon lights popped
and blinked 'Time expired',
as on a parking meter.

Mbuyiseni Oswald Mtshali

Summer full moon

The cloud tonight
is like a white
  Persian cat –

It lies among the stars
with eyes almost shut,
lapping the milk from
the moon's brimming dish.

James Kirkup

Quiet secret

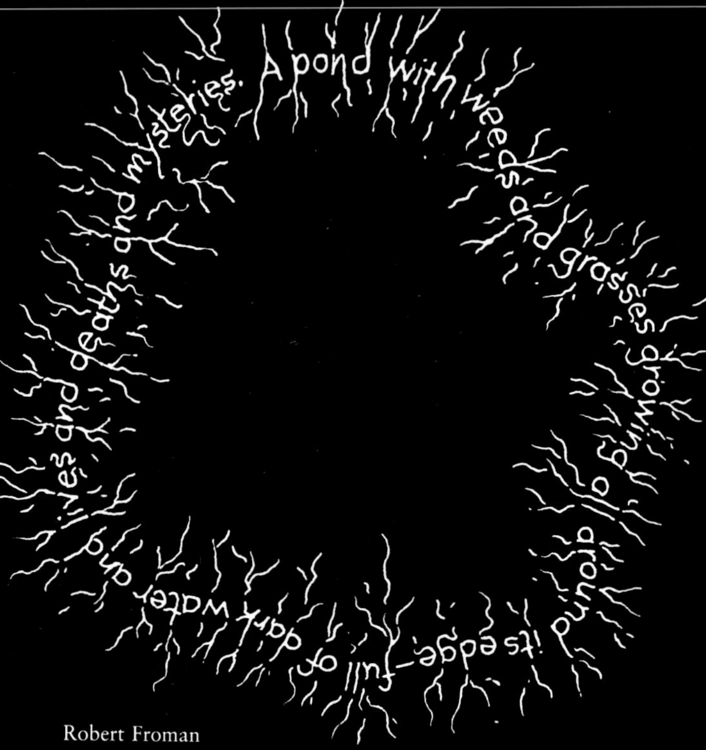

Robert Froman

The cornfield den

The boy pointed
And his fingers said,
'Step, silly, in between the rows;
Mind how you tread;
Don't crush the wheat,
Go in the field ten yards or so,
Then on your knees,
And start to beat
A narrow track.
Get down and kneel;
I'm coming in behind,
Now crawl,
And keep your head down low.'

They brushed aside the corn
That creaked and hissed;
With careful foot
And balanced step
They pressed between
The heavy-headed grain;
Then down on knee,
Elbow and thigh,
Out of the meadow green darkened by hedge and tree
Into the brittle forest of dusty gold.
The wheat was strong and high
As the girl's throat,
High as his jersied chest;
The cable-knitted ears
Scraped at the blue wash of the sky.

'Lower your head
And mind your eye.'
As he spoke
He showed how to use elbows.
The stalks splintered and broke.
Between their parted teeth the children hissed,
Sucked and grunted at the pain,
And scratched by spiky straw beneath
They gasped along the flattened grain.

Deep in the middle of the field
They hollowed out a lair;
They lay and squeezed their bloodied knees,
And in the pollen-dusty air
Palmed up their nose to squash a sneeze.

A scrape of tyres, an engine roar,
The grind of car in bottom gear,
Then silence.
Inside the den they shrank and felt
The sweet uncertainty of fear.

The farmer scanned the lane, the field,
The heifers bunched beside the hedge,
His ricks, the fold-yard and the pool
Ringed with the velvet cover of the sedge.

Deep in the restless waves of wheat
The children cowered.
'The dratted kids; somewhere around.'
Only the swell of corded grain
And two heads pressed against the ground.

The farmer sniffed and thumbed his pipe,
Drove on along the rutted track.
'Old Fisher's gone.'
The boy sprawled idly on his back;
The girl flung back her hair
And for a moment tried to stare
The blue tent of the sky
Which with a summer brilliance burned
Too bright for the unshaded eye.

How good it was to lie and hide
And dream,
Cheek on your hand,
Of that imaginary land,
Deep in the wheat,
Where possibly a knight
Was wandering with a blazoned shield;
Perhaps a snorting dragon forced to yield,
Or maybe an explorer and his guide,
Perhaps...

The children lay
And breathed in deep and slow
The long and leisurely delight
Of quiet, unhurried holiday;
While round their heads upon the ground
The mystery and imminence
Of territory unexplored;
The gold sun poured
Its warmth upon their secret hide;
And, closely ranked, the forests stride
Illimitable and immense.

He stirred.
'Come on; wake up.'
She hugged her knee;
'I bet old Fisher's gone for tea.'

Gregory Harrison

We say

you must climb, for this dare
up the high iron railings with spikes on, *here*,
by the great black gates, pulled together with padlocks and chains.
Take hold of the branch of the elm that bends
as you go out along it, hand over hand
hanging down
as it lowers you in
till you drop
like a ripe apple and plop
into grass that has never been mown.
Then you jump
up, and run doubled for shelter, to the clump
of dark rhododendrons – push through
the shrubs that snatch at you;
hollies as spikey as metal
ground-elder and dead-looking laurel
that turns to a tunnel...
... you crawl
in a funny cat-smell, right up to the wall
where the rose sags away from the stone;
under the high barred windows that stare
like bits of the sky, from the big house where –
so they say –
down at the end of a long corridor,
the rich man lives, in one room, on his own.

We say you can hear:
the wind pushing under the door,
the chiming of clock after clock
the creak of a stair, or the click
as the key turns in a lock –
he is going from one room to another
where dust-sheets still shroud all the furniture...

– Maisie says, someone looked through the letter-box once
and a pair of pale eyes peered into their own
then three fingers came through like a tongue from a mouth
along with a laugh –

No one's seen all of him, but
if he sees *you*, he comes running out!

... But when I
shot out of the shrubs, up the steps, rang the bell
and turned, and jumped to the gravel
running as though I was just out of Hell,
down the moss-softened drive to the gates
and I fell –
nothing happened at all

so I stopped:

And there at a window, high up,
he was standing, quite still
in bow-tie and tails, like someone who's off to a ball,
watching –
as if he had been there for ever.

– It was just as if I'd been caught,
but nothing would happen,
nothing would happen at all.

Brian Lee

Next door

My mum says
the woman next door
isn't a fly,

a huge bluebottle
rubbing six thin legs together
crawling upside-down on the ceiling
sticking her long nose into the jam.

My mum says
that buzzing and whirring and humming
we hear each day through the wall
is only a hoover.

If that's true, why
does her husband scuttle
over the floor on eight hairy legs
and build thick webs
in the dark cupboard under the stairs,

and why does Stan
her eldest son,
buy huge cans of Deadly Flykill?

When I next see her
zooming over the compost and dustbins
I'll have to ask her
just what's SWAT.

David Harmer

Conversation with insults

'Who do you think you're staring at then?'
'You. You look mad in those specs.'
'Not half as mad as you look, my son –
And when are you going to wash next?
You smell, you're disgusting, you need a good bath.
And why do you wear those daft clothes?'
'You're an idiot, you're brainless, you're grotty as well.
Get a hanky and clean up your nose.'
'I'll clean up *your* nose with my fist in a minute
As a favour to everybody.
You're due for a thumping, it's just what you need – '
'A thumping? You couldn't thump Noddy!'
'You couldn't thump Big Ears – you've got them as well.'
'You're a creep!'
'You're a louse!'
'You're a snake!'
'You're a moron, a cretin – hey, there goes the bell!'
'I enjoyed that. See you next break.'

Eric Finney

Sarky devil

Our teacher's all right, really,
But he can't stop making
These sarcastic remarks.

Nev Stephens, who's got no one to get him up,
Creeps in fifteen minutes late
Practically every morning –
Always with a new excuse.
And old Sarky says,
'Good evening, Stephens,
And what little piece of fiction
Have you got for us today?'

He even sarks the clever kids:
Says things like
'Proper little Brain of Britain, aren't we?'
And only yesterday when Maureen
(Who's brilliant at everything)
Complained of a headache,
He says, 'Perhaps it's the halo
Pinching a bit, Maureen.'

And then there's Bill Nelson
Who's just one of these naturally scruffy kids
Who can't keep anything clean –
Well, his homework book's a disgrace,
And Sarky holds it up
Delicately, by a corner,
At arm's length, and he says,
'Well, Lord Horatio,
Did you have your breakfast off this
Before or *after* the dog had a chew at it?'

Even the one who's getting sarked
Laughs –
Can't do much else, really;
And the rest of the class
Roars, of course,
And feels like one big creep.

It's a good job we understand
Old Sarky.

Eric Finney

A chemistry student from Gillingham

A chemistry student from Gillingham
Kept emptying jam jars and filling 'em
  With a poisonous jelly
  That was bright green and smelly
So he used it on teachers for killing 'em.

John Rice

Glenis

The teacher says:

Why is it, Glenis,
Please answer me this,
The only time
You ever stop talking in class
Is if I ask you
Where's the Khyber Pass?
Or when was the Battle of Waterloo?
Or what is nine times three?
Or how do you spell
Mississippi?
Why is it, Glenis,
The only time you are silent
Is when I ask you a question?

And Glenis says:

Allen Ahlberg

Poem about writing a poem

'Write a poem,' she says
'About anything you like.'
You can practically feel the class all thinking,
'On your blooming bike!'
A poem! I'll tell you one thing:
Mine's not going to rhyme.
A poem between now and playtime!
There's not the time.
In half an hour she'll say,
'Have you done? Hand papers in
And go out.'
I mean, does she have the slightest idea
What writing a poem's about?
I mean, it's agony:
It's scribbling thoughts
And looking for rhymes
And ways to end and begin;
And giving it up in total despair –
'I'm chucking it in the bin.'
But tomorrow it pulls you back again,
And hey, a bit of it clicks!
And you sweat with the words
But it's hopeless again
And it sticks.
And you put it away for ever...

But it nags away in the back of your head
And the bits of it buzz and roam,
And maybe – about a century later –
You've got a kind of a poem.

Eric Finney

Poetry (or cricket)

Writing poetry is like fielding
at cricket. You stand on the grass
watching, ready to act.
If you're in the right place
at the right time and looking
in the right direction
you can hope to contribute,
occasionally, something useful
to the game.

And there's always a hope of glory;
a moment may offer itself when everything
conspires in your favour –
the light, the wind, the batsman
(Whoever he may be).
A beautifully lobbed ball
will seem to slow down in the air
and curve its parabola
right into your waiting hands.

Pamela Gillilan

The sum

He can't get it right &
can't get it right no matter how

hard he tries no matter he knows
the answer he *must* show

the working out how he arrived
at it his teacher wants
to see

how many bricks
& which way up & how many ways
into this big box but

they won't fit & nothing will
make his tears come
a bit more slowly

he cries on the bus home
& all evening until
the sum sends him to bed numb

a bit stunned until
sleep rubs out the question

& bricks are smallest parts of
smallest particles

& night fills up his box

Les Arnold

Ars mathematica

Any
triangle
with two equal angles
of however many degrees you please
will be known by the grand name of isosceles.

Parallelogram
not a problem
a rectangular
dipsomaniacal
quadrilateral.
Right angles?
Not necessary.
Systematical!
So ecstatical
problematical
box that I am,
Parallelogram!

A
square
is four-sided
completely right-angled
nothing new-fangled
or rare in a
square

A trapezium
Here is an easy one
two sides are parallel
this you can follow well
two more can be symmetrical
result can even be poetical!

Next
the hexagon
Here's a pesky one
six axes of symmetry
so handy in geometry
tessellates simply
polygon for
six

You have certainly seen a rectangle
if you've ever been in a quadrangle
four sides and equal sides opposite
four right angles, not my favourite

Judith Nicholls

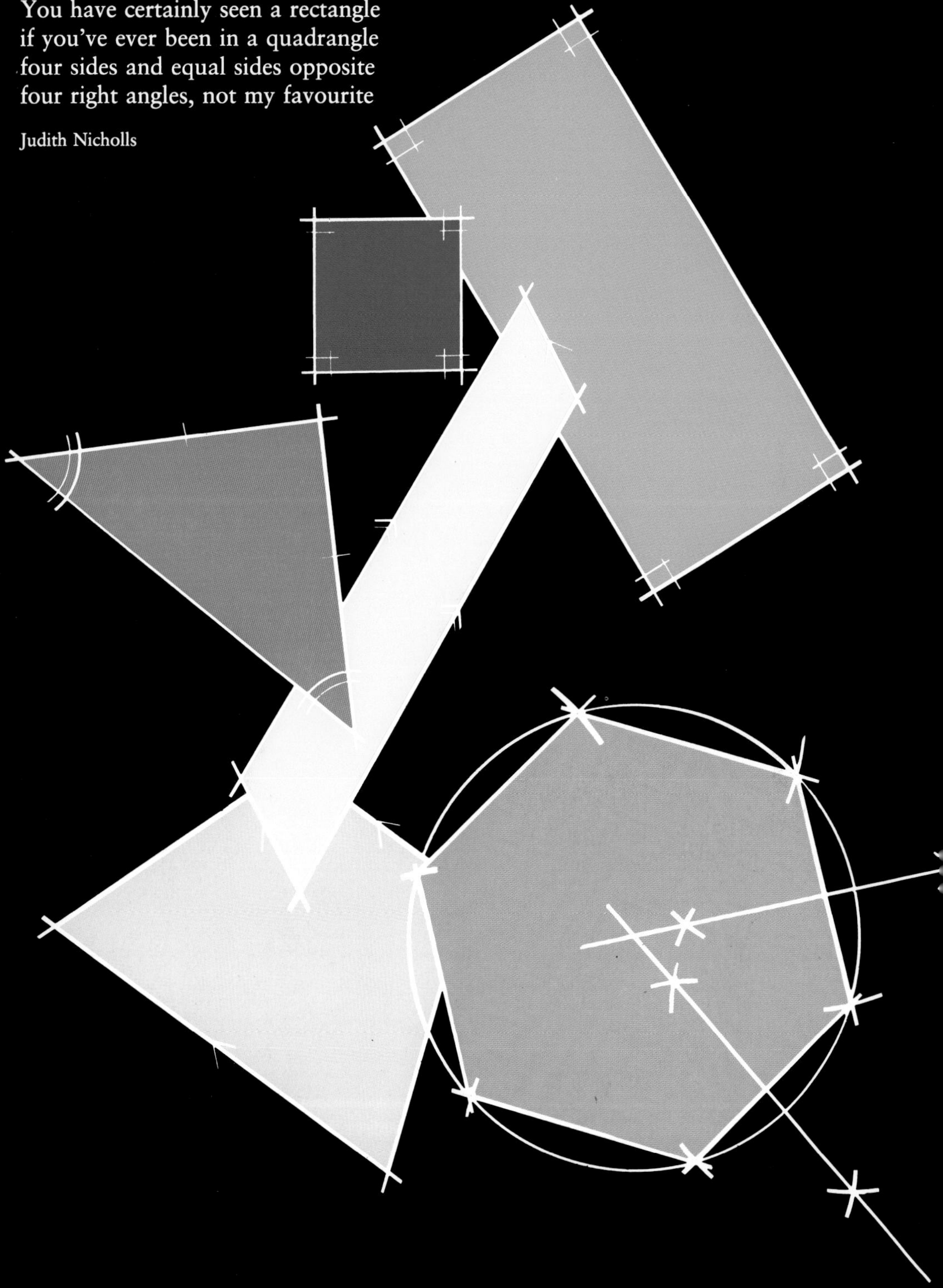

Grown up?

When I'm grown up?
What do I want to be?
Well, Sir,
Since you ask,
I wouldn't mind...
I wouldn't mind being a *tree*!
I'd like to push my feet
Into the soil,
And stand there
A few hundred years,
Just *being*...
Just being a tree.

But if I can't be a tree,
I think I'd just like to be *bad*.
I'd like to have cakes and champagne in the afternoon,
And eat condensed milk straight out of the tin,
With a green plastic spoon;
I'd have handmaidens bringing sherbet,
And a pet boa-constrictor called Herbert.
I'd lie on a silken divan,
Reading lurid novels;
And make fat profits,
From people in hovels;
I'd go to Oxford,
Take my teddy-bear,
And dine on oysters and paté,
Gin and jugged-hare.
I'd...
But he's not listening,
They never do...

One thing I'll *never* be,
When I grow up;
I'll *never* be,
A fat man,
In a pin-stripe suit,
Who smiles his fat smile,
At a pimply Second-Year,
Lays a heavy hand
On the boy's shoulder,
And says,

'Now, young man,
What do you want to be,
When you grow up?'

Then walks away,
Without listening,
With a sneer and a wink
At the Head,
And a burst of coarse laughter,
Down the corridor.

John Cunliffe

Four o'clock Friday

Four o'clock, Friday, I'm home at last
Time to forget the week that has passed.

On Monday, at break, they stole my ball
And threw it over the playground wall.

On Tuesday morning, I came in late,
But they were waiting behind the gate.

On Wednesday afternoon, in games
They threw mud at me and called me names.

Yesterday, they laughed after the test
'Cause my marks were lower than the rest.

Today, they trampled my books on the floor
And I was kept in, because I swore.

Four o'clock, Friday, at last I'm free.
For two whole days they can't get at me.

John Foster

Pocket money

'I can't explain what happens to my cash.'
I can, but can't – not to my Mum and Dad.
'Give us ten pee or get another bash' –

That's where it goes. And though their questions crash
Like blows, and though they're getting mad,
I can't explain what happens to my cash;

How can I tell the truth? I just rehash
Old lies. The others have and I'm the had:
'Give us ten pee or get another bash.'

'For dinner Dad? ... just sausages and mash.'
'That shouldn't make you broke by Wednesday, lad.'
I can't explain. What happens to my cash –

My friends all help themselves. I get the ash
Of fags I buy and give, get none. 'Too bad.
Give us ten pee or get another bash

For being You.' And still I feel the thrash
Of stronger, firmer hands than mine. The sad
Disgust of living like a piece of trash.
I can't explain what happens to my cash.
'Give us ten pee or get another bash.'

Mick Gowar

One parent family

I wish that I had more than just
A weekend dad, a dad who shakes
The dust of Sunday from his feet
And makes his short drive home
When we've been sourly dumped
At mother's peeling five-day door.
It isn't only childish greed
That makes me want my father more.

John Kitching

There are four chairs round the table

There are four chairs round the table,
Where we sit down for our tea.
But now we only set places
For Mum, for Terry and me.

We don't chatter any more
About what we did in the day.
Terry and I eat quickly,
Then we both go out to play.

Mum doesn't smile like she used to.
Often, she just sits and sighs.
Sometimes, I know from the smudges,
That while we are out she cries.

John Foster

Billy and me and the pane

'That's the fourth window you've broken this year.'
The voice is grim – it's Dad's.
'This time you fix it yourselves –
I mean it, my lucky lads.
Get the rest of the glass out,
Clean out the rebate, measure it.
Order the new pane from your Uncle Ted.
Oh, better not forget the putty either,
There won't be enough in the shed.
Then you put the new pane in
Neatly – and I do mean neatly.
You do it all, the whole shebang,
All by yourselves – completely,
I mean it.'

He means it.
'Hey, what's the rebate?'
'I dunno. What's the shebang?'

Three whole days it took us to fix
That ruddy window pane:
Three days out of our summer holidays,
Three sweltering hot days
Of worry and work and strain –
No swimming, no fishing, no football,
No daily bike-ride;
Just pestering and worrying
About fixing that window
Seventy-three centimetres long and
A fraction less than
Forty-seven and a half centimetres wide.

Billy cut his hand twice and
I cut an arm and an ear
Getting the spikes of glass out.
Then that old putty was
As hard as granite –
We ruined two of Mum's
Best kitchen knives on it.
We nearly gave up in despair
Until Dad suggested a hammer
And an old chisel.
Mind you, he didn't lay a finger on the job:
Just smiled, like a cross between
Dracula and a Cheshire cat.

One whole day and about
Twenty-three million hammer taps later
We'd cleared out all the spaces.
We'd also cobbled and splintered the wood
In about a thousand places.
'Don't worry,' said Billy with a sickly grin,
'It won't show when we whack the putty in.'

This Uncle Ted in the shop:
He's not really an uncle at all,
Just a boozing pal of Dad's.
He kept us waiting for ages next morning
Until his boss said, 'Ted, it's time
You served those lads.'
Even then he did his best to defeat us:
Refused to have anything at all
To do with centimetres.

'You come back with some sensible measurements,'
He gruffs,
'That's when the glass-cutting starts.'
So we go home and measure it again in inches
And we're off down to the shop again
With rage and hate in our hearts.
'Come back at three o'clock
And it might be ready,' says
Uncle Misery Guts.
It wasn't, of course, nor at four o'clock,
And at five the blooming shop shuts.
But just before closing-time we did finally
Get a pane of glass and
A tub of putty out of Uncle Sunny.
At a price:
Three weeks' pocket-money.

But there's a spring in the step now
And a bit of cheerfulness creeping back
Into Billy and me:
It's nearly fixed! Visions of
Lazy days and Dad back to normal
Instead of Mr Misery.
Guess what?
THE GLASS DIDN'T FIT –
Too big by the tiniest bit.

'Let's leave home,' I said and
Billy agreed –
We'd take off after tea:
Grab 'jamas, toothbrushes, football gear
And head for the hills or the sea.

We'd leave a note saying
'Gone for ever.
You'll never find where we hide.
And this is all because of that pane of glass:
We failed – but at least we tried.'
They'd find the note at breakfast time –
Imagine Dad's sorrow and shame . . .
But Billy said, 'Hey, if we run away
We'll miss Saturday's Rangers' game.'
So we called it off, and then Dad turned up
And grinned to see we were stuck.
He looks at the problem and says,
'Well now chaps, I think with a touch of luck
You'll get it to fit by chiselling a bit
Off the wooden jamb.'
And after about another two hours' work
It did.
Shazam!

Dad also said, 'That putty's no good,
It's for metal frame windows.
You want putty for wood.'

After tea, Billy heard Mum
Saying to Dad,
'Oh, give them a hand now, Dan.
They've done ever so well.'
Dad said nothing.
Hard man.

I always thought putty was easy –
Like plasticene,
But it's more like that gluey school pud,
And on day three we got into a right mess
With plenty of putty sticking to our hands
And the knife, but not all that keen
To stick to the wood.
We got the hang of it in the end, though:
Rolled thin sausages of putty and pressed
Them into the rebate. Good!
Great moment: the pane is
Lifted into place and pressed firmly
Into the frame of putty –
Magic! It's in!
Knock off for a large jam butty,
Then back to finish the job which
Now is going a dream and
Almost seems like fun.
O.K., so the window's covered with
Greasy finger-prints and it's
Impossible to get the outside putty smooth,
But it's done!

Mum produces a large cold bottle of lemonade
For lunch.
'To celebrate,' she says. 'Champagne for three.'
'I don't know about *sham*,' says Bill,
'Seemed like three days of *real* pain to me.'
And Mum laughed something silly:
He's a proper wit that Billy.

'Not bad,' Dad says when he gets home
From work. 'Putty's a bit lumpy but
It's not too much of a hash.'
And he takes the knife and zips
It rapidly round the four edges of putty
Levelling them all in a flash.
'Just a little knack,' he smirks,
'Took you three days I reckon that job did ...
Three days work for two men ...
A reasonable job ... here, share this quid.
And in future you might take a bit more care
When you're belting that football everywhere.'

And he could be right, the miserable beast.
We might.
For a bit, at least.

Eric Finney

Guardian

Sprawled across the stair
the cat blocks my way.

Shoo! I nearly say
but the sun in which

he basks gilds the rich
colours of the stained

glass window behind
him and the brief gleam

has briefly turned him
into a tiger.

Keith Bosley

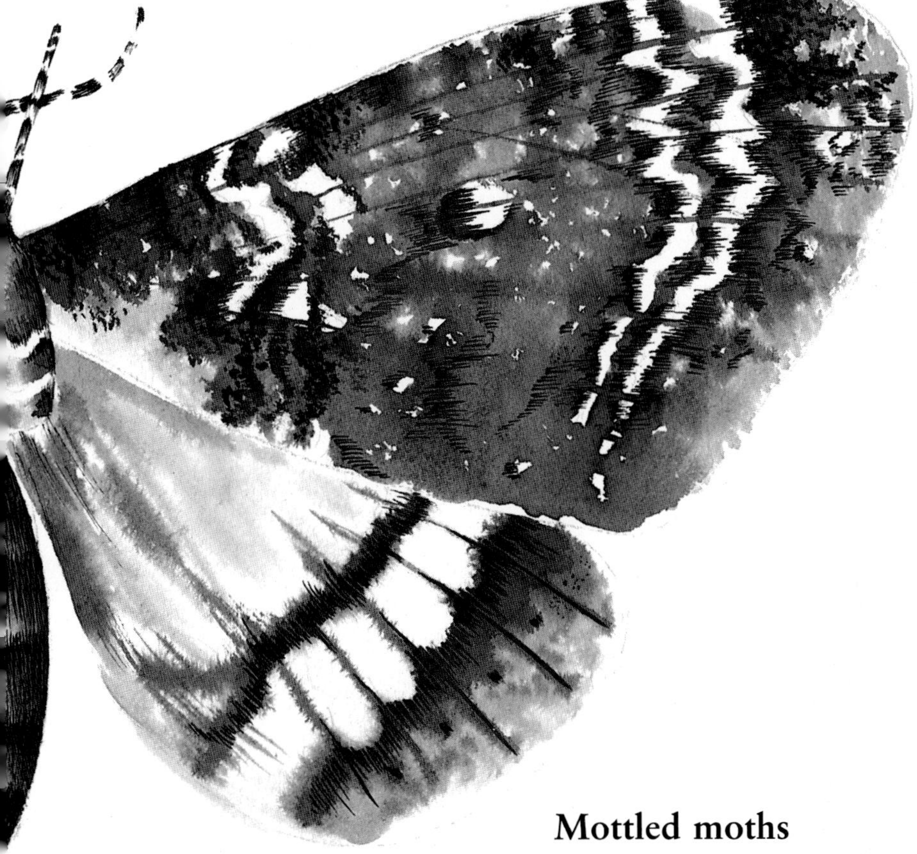

Mottled moths

Sometimes they settle on a white wall
and then you see them, soft unwarlike
arrowheads of tortoise-shell,
cream and brown.

But they are tree bark
or old stone, and can stay safe
and invisible all day
on the unsheltered rock,
the wild tree.

They sleep and bask hidden
by their own openness,
though birds that are always hungry
live on the same branch.

Pamela Gillilan

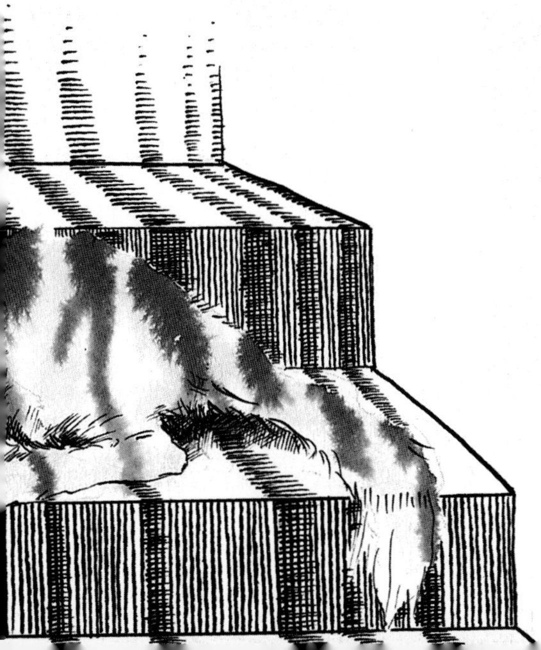

On a blue day

On a blue day
when the brown heat
scorches the grass
and stings my legs with sweat

I go running like a fool
up the hill towards the trees
and my heart beats loudly,
like a kettle boiling dry.

I need a bucket the size of the sky
filled with cool, cascading water.

At evening
the cool air rubs my back,
I listen to the bees
working for their honey

and the sunset pours light
over my head like a waterfall.

David Harmer

Autumn

Autumn is the rich season,
the year's tycoon;
great armfuls of
the trees' lost
bullion;
cascades of newly
minted gold;
the season's banknotes –
crinkled, crisp, already
spent.

Adrian Rumble

on the doorstep

on the doorstep
in a brief autumn sun
a beetle z
i
g
z
a
g
z
i
g
z
a
g
s by
its black back blazing

John Rice

Haiku for autumn

Hedgerows fired with beech,
lawn smouldering with apples;
golden October!

Silence pungent as
oranges, darkened summits
humped with wind-bared rock.

Rush of water over stone
and not-quite-silences
of wind through trees.

Judith Nicholls

West wind

roused us last night
with his roistering clatter
rudely bursting through
latched windows
locked doors
without so much as a
'May I come in?'
or ring of the
midnight bell

lifts curtains
rushes impatient
under doors
swirls down chimneys
like some drunken
blustery Santa
lost in time.

Judith Nicholls

Still

Snow, finer than crushed glass,
Falls continuously,
Smothering road and roof
Until the day shivers
To a Christmas card still.

Buried, the earth's slowed pulse
Weathers this deep season.

Wes Magee

School field in winter

Hundreds of Wellington boots
Have hard-packed the snow until
The field glitters like a rink.

Snowmen stand rigid with cold.
Squeals of children skate across
An acre of Arctic ice.

Wes Magee

Christmas landscape

Tonight the wind gnaws
With teeth of glass,
The jackdaw shivers
In caged branches of iron,
The stars have talons.

There is hunger in the mouth
Of vole and badger,
Silver agonies of breath
In the nostril of the fox.

Laurie Lee

Xmas

Not a twig stirs. The frost-bitten garden
Huddles under a heaped duvet of snow.
Pond, tree, sky and street are granite with cold.

In the house, electronic games warble:
Holly awaits the advent of balloons,
And the TV set glows tipsy with joy.

Wes Magee

Doctor Christmas

My first thought: 'Have they sterilised the beard
Worn by so many Father Christmases
Before me?' Then: 'These children, they've just heard
My voice, their doctor's voice, on a ward round,
And seen my smile; they'll notice who it is
Under the outfit when I come again
As Father Christmas. Well, here goes . . .' No sound
Escapes my lips at the first bed; small fingers
Reach for the offered joy; each counterpane
Becomes a magic carpet. A fat boy
Gives me a wink: 'I've seen you!' His glance lingers.
'Today?' I ask – but that's not what he meant;
'No; last year, at Christmas'. Grabbing his toy,
He thanks and greets me as a long-lost saint.

Edward Lowbury

Santa

Do you believe in fair play?
Liberty? Equality?
Then you may as well believe
In Santa Claus: each is a myth,
But each lives on from year to year.
So, with the carols, lure him back,
The bearded gentleman in red
Who brings a bonus of free goods;
Lover of kids and animals.

We hear the sleigh-bells in our sleep,
Tear open wrappings before dawn –
But never stop to question where
He gets those goodies. If the myth
Is not to perish, don't look now
Or you may catch him, wearing mufti
In your larder, making off
With something you won't miss just yet;
But if you see him filch the stuff
You will no longer find it strange,
This flouting of terrestrial laws:
Shake 'Santa', and the name will change
To 'Satan'; 'Claus' will change to 'Claws'.

Edward Lowbury

Muntze of the ear

Jan you are Ray in disguise.
Fair brewery does not always mean fair beer.
Marge is easier to spread than butter.
Ape reel is not a film about King Kong.
Mare is pronounced the same as 'mayor'.
Dune is where sand grains gang up on the sea.
Jewel eye doesn't mean see-through diamonds.
A gust of wind blew right through my ears.
Accept timber if you ordered bricks but only wood was available.
Och Tober is a Scottish way of saying 'Oh dash!'
No Farnborough airshow this year due to Jumbo on the runway.
Does amber mean stop on traffic lights?

John Rice

The calendar of my year

JABUARY	FEVERARY	MARSH
HAYDRILL	PERHAPS	SOON
DEW LIE	HAW GUST	SOFT UMBER
OCTOBURR	NEW EMBER	DISMEMBER

Geoffrey Summerfield

Haiku calendar

Leaf, brown in the wind
Taps like a bird on the pane
Presaging Autumn.

Leaf, sunk in the soil,
Spends winter dying; new life
Mounts from its leaching.

Leaf, loosing its sheath
Into the warm air unfurls
Green in Spring's morning.

Leaf drinks air and light,
Channels rain, feeds its parent
Great tree of Summer.

Leaf, brown in the wind
Taps like a bird on the pane
Presaging Autumn.

Pamela Gillilan

The trees

The trees are coming into leaf
Like something almost being said;
The recent buds relax and spread,
Their greenness is a kind of grief.

Is it that they are born again
And we grow old? No, they die too.
Their yearly trick of looking new
Is written down in rings of grain.

Yet still the unresting castles thresh
In fullgrown thickness every May.
Last year is dead, they seem to say,
Begin afresh, afresh, afresh.

Philip Larkin

April birthday

When your birthday brings the world under your window
   And the song-thrush sings wet-throated in the dew
And aconite and primrose are unsticking the wrappers
   Of the package that has come today for you.

      Lambs bounce out and stand astonished
      Puss willow pushes among bare branches
      Sooty hawthorns shiver into emerald

                                    And a new air
      Nuzzles the sugary
      Buds of the chestnut. A groundswell and a stir
      Billows the silvered
      Violet silks
      Of the south – a tenderness
      Lifting through all the
      Gently-breasted
      Counties of England.

When the swallow snips the string that holds the world in
   And the ring-dove claps and nearly loops the loop
You just can't count everything that follows in a tumble
   Like a whole circus tumbling through a hoop

      Grass in a mesh of all flowers floundering
      Sizzling leaves and blossoms bombing
      Nestlings hissing and groggy-legged insects

                                    And the trees
      Stagger, they stronger
      Brace their boles and biceps under
      The load of gift. And the hills float
      Light as bubble glass
      On the smoke-blue evening

And rabbits are bobbing everywhere, and a thrush
Rings coolly in a far corner. A shiver of green
Strokes the darkening slope as the land
Begins her labour.

Ted Hughes

The half-mile

I was twelve when I swam the half-mile,
up and down the tide-fed cold concrete
pool, with a slow steady side-stroke.
My father counted the lengths,
at first from the deep-end board
and then, as I moved more laboriously,
pacing alongside, urging me on.

The race was only against myself
and distance. The grainy salt water,
though not translucent like the chlorinated
blue lidos of town, buoyed me helpfully,
lapped softly against the bath's grey sides
variegated with embedded hardcore pebbles.
I swam from goal to alternate goal; he counted.

When he called enough I scrambled
over the sharp shutter-cast lip,
shuddered into a dry towel, drank
the words of praise. The planks
of the changing-room walls
were warm to touch. It had seemed to be
a great deal of swimming; still does.

Pamela Gillilan

Racing cyclist

His feet clipped to it, he turns the treadmill
Of his double chain wheel, in highest gear.
The early morning mist on the level road
Through a low-lying countryside
Retreats before him, dragging its cloak
Over the hedges, the lines of poplar trees
And towns and villages with cheering crowds,
The sun, like everyone else, coming out to watch.

For miles, he himself and the riders beside him
Seem to him to be standing still,
All moving at the same high speed.

Under welcoming banners and past advertisements,
Low on the handlebars, he ducks the air
That blocks his way and clutches at his clothes,
Keeping level above all with himself
And not a second behind the best he can do.

He rides as surely as if his narrow tyres
Fitted into a groove already there
Or followed a chalk line drawn to the finish
Where people leap up at the roadside,
Beckoning and calling a winner out of the pack.

Stanley Cook

Cycle song

Tonight the cousin I grew up with
goes to Australia.
As children we rode bikes down
Yorkshire summers, our brown knees
clipping the handlebars.

In four days he will be lowered
into the cool of a Brisbane bungalow
he will get fat and begin to turn
into a faded photograph
I will never see him again.

Yesterday I noticed that the rooks
had gone, my father said they hadn't
been back in two years.
The nests are turning to wire wool
snatching a living from branches.

Now I think about it the old man
who pushes his wheelchair along
our street hasn't been seen for weeks.
Someone has folded him up with the chair
he is almost forgotten.

It's the constant erosion that's hardest
to take, the relentless pounding of
rocks into sand.
Soon I will be sand, I will be in a
crowd pushing wheelchairs heavy with
rooks through the cool dark of
a bungalow in Brisbane.
My cousin will watch from his dusty
print on the wall.

Martyn Wiley

Living on

My mother died eight years ago
When I was five, but still awake,
Asleep, I see her summer face
And feel her living hand on mine.
No ghost, but there and meeting need
And helping me to know there are no dead.

John Kitching

Uncle Stan

Here's Uncle Stan, his hair a comber, slick
In his Sundays, buttoning a laugh;
Gazing, sweet-chestnut eyed, out of a thick
Ship's biscuit of a studio photograph.
He's Uncle Stan, the darling of our clan,
Throttled by celluloid: the slow-worm thin
Tie, the dandy's rose, Kirk Douglas chin
Hatched on the card in various shades of tan.

He died when I was in my pram; became
The hero of my child's mythology.
Youngest of seven, gave six of us his name
If not his looks, and gradually he
Was Ulysses, Jack Marvel, Amyas Leigh.
Before the Kaiser's war, crossed the grave sea
And to my mother wrote home forest tales
In Church School script of bears and waterfalls.

I heard, a hundred times, of how and when
The blacksmith came and nipped off every curl
('So that he don't look too much like a girl')
And how Stan tried to stick them on again.
As quavering children, how they dragged to feed
The thudding pig; balanced on the sty-beams,
Hurled bucket, peelings on its pitching head –
Fled, twice a day, from its enormous screams.

I watched the tears jerk on my mother's cheek
For his birth day; and gently would she speak
Of how time never told the way to quell
The brisk pain of their whistle-stop farewell:
A London train paused in the winter-bleak
Of Teignmouth. To his older friend said, 'Take
Good care of him.' Sensed, from a hedging eye,
All that was said when neither made reply.

I look at the last photograph. He stands
In wrinkled khaki, firm as Hercules,
Pillars of legs apart, and in his hands
A cane; defying the cold lens to ease

Forward an inch. Here's Uncle Stan, still game,
As Private, 1st Canadians, trimmed for war.
Died at Prince Rupert, B.C. And whose name
Lives on, in confident brass, for evermore.

That's all I know of Uncle Stan. Those who
Could tell the rest are flakes of ash, lie deep
As Cornish tin, or flatfish. 'Sweet as dew,'
They said. Yet – what else made them keep
His memory fresh as a young tree? Perhaps
The lure of eyes, quick with large love, is clue
To what I'll never know, and the bruised maps
Of other hearts will never lead me to.
He might have been a farmer; swallowed mud
At Vimy, Cambrai; smiling, have rehearsed
To us the silent history of his blood:
But a Canadian winter got him first.

Charles Causley

Who is you?

Let's all climb the family tree
See who's you and who is me.

If you are your brother's brother,
Your grandad's daughter is your mother.
But while we're finding who is who
Your grandad's little girl's little girl's you.

If you're a girl climbing family trees,
Your father's brother calls you niece.
He calls you nephew if you're a lad,
But says 'Hi brother!' to your dad.

Now the family tree is buzzing,
Your brother's sister's kid's your cousin.
Your nephew's uncle's you you see,
And my father's brother's nephew's me

The more we climb the more the fun,
Your great grandma's grandad's mum.
And here's something else you'll find is true,
Your sister's baby's uncle's you.

Up the tree, oh this is the life,
Grandma's daughter-in-law's uncle's wife.
Let's all climb the family tree,
If you're your mum's only child, you're me.

Rony Robinson

Aunty Joan

When Aunty Joan became a bone,
The family didn't applaud her.
Uncle Ted just shook his head.
The rest of us *ignawed* her!

John Foster

My Uncle Justin

His father's name was Henry Symes.
When Justin thought of the old man's many crimes –
Not least of which was siring him, Justin –
And fearing that there might be something in
The biblical threat of the father's sin
Being laid at the son's door with dire penalties to face,
He decided to change his name from Symes to Case.
Justin Case.

Vernon Scannell

Stage-struck

'Well then, young Darryl' said Uncle Tom,
'What are you gonna be when you grow up?'
'An actor.' 'An akhtoor. That's right champion.'
Uncle Tom struck himself in the gut.

'Did I ever tell you, young Darryl,' he mouthed
'How I once played York in Richard the Second
For the Batley Town Council Players. Pow!
I went through that audience like a Black-an'-Decker.

'But you, young Darryl, you'll want to play Richard,
Richard himself, then Hamlet, then Lear –
The climax of every actor's ambition.
I'm sure you'll move your audience to tears.

'I'm right glad, Darryl. So few young chaps
Are moved by the ring of Shakespeare's lines.
"To be or not to be ..." eh lad, that's
The stuff to drive you out of your mind.'

'Well uncle, I wasn't thinking of Shakespeare,
More like of being a teevee star,
And making commercials in Las Vegas
For Flora and Mannikin Cigars.'

Leo Aylen

Uncle William

I stayed with you once
in your tiny church-lane cottage
with the outside pump, the velvet cloth
and sing-songs cramped around the piano.

With black-fringed stumps of fingers,
braces, ample paunch,
you could have been
miner, dustman, sweep –
but no; village blacksmith
fitted best that village scene.

I remember strong green soap,
tin bowls of icy water for the morning wash;
my aunt's night-calling for the cat
across still hedgerows and the cobbled lane,
a shared bed with spoiling cousins,
Billy Bunter by oil lamp at forbidden hours
and orange moths against the darkened pane.

Uncle William. Dead now;
the blacksmith and the cottage gone.
No cobbled lane but just a road now,
a road my aunt must tread alone.

Judith Nicholls

Grandparents

They love their
grandchildren

They advise, they
give

The stories of the past
they tell

They teach
you

They're a part of
your family

They are not old-
fashioned, it's just
their views.

They created you!

You will be one

If you destroy them
you destroy a generation –
your creators.

(this voice is of a computer)

A waste of space

Use a computer

Bad old news think
ahead

Go to school

They waste food and
occupy space.

Their brains are slow-
they have no use to
the State, they should
be exterminated...

insignificant data...

insignificant data
insignificant data

This cannot be
classified – they create
germs – a waste of space...
... This cannot be processed.

Maxine Kelly-Robinson

Great-Gran

Great-Gran just sits
All day long there,
Beside the fire,
Propped in her chair.

Sometimes she mumbles
Or gives a shout,
But we can't tell
What it's about.

Great-Gran just sits
All day long there.
Her face is blank,
An empty stare.

When anyone speaks,
What does she hear?
When Great-Gran starts,
What does she fear?

How can we tell?
For we can't find
A key which can
Unlock her mind.

Great-Gran just sits,
Almost alone,
In some dream world
All of her own.

But when Mum bends
Tucking her rug,
Perhaps she senses
That loving hug.

John Foster

Leave-taking

The only joy
of his old age
he often said
was his grandson

Their friendship
straddled
eight decades
three generations

They laughed, played
quarrelled, embraced
watched television together
and while the rest had
little to say to the old man
the little fellow was
a fountain of endless chatter

When death rattled
the gate at five
one Sunday morning
took the old man away
others trumpeted their
grief in loud sobs
and lachrymose blubber

He never shed a tear
just waved one of his
small inimitable goodbyes
to his grandfather
and was sad the old man
could not return his gesture.

Cecil Rajendra

Grandma

Up at dawn ...
While grandad slept
Off last night's beer,
Lighting fires in the
Black-leaded grate,
Drawing water from the well
In a bucket of wood
On a rope,
Feeding the chickens,
Collecting eggs ...
Putting a china egg
In my very own eggcup
To tease me ... every year.
Grandma, with rosy apple cheeks
And snow-white hair, speaking the
Rolling tongue of Gloucestershire.
Sleeves rolled up, and ice-cold water
In a bowl on the low wall, where
I sat and shivered in anticipation
Dressed in my school knickers ...
Watching her deft, plump hands plunge
To catch the red carbolic soap
And scrub my face.
Oh, the breathless sudding, dare I yell
In protest and the soap was thrust
Into my gaping mouth,
To cling its taste to my teeth
All day long.
Grandma, who dug the garden
And cooked the dinner,
Called grandad from the pub,
Collected the fruit ... and loved me.
Grandma, who threw back the frogs
She drew with the water,
Never needing to raise her voice ...
Just one look and mischief died
A bit ashamed ...
Undemonstrative, hard-working, rough.
Forest born, bred and buried.
Oh, how I loved my grandma ...

Joan Batchelor

Then

They never expected it of my grandmother,
all this choice.
Stolid, vocationally-trained
with neat samplers and clear instructions on
pastry-making and how to preserve the strawberries,
for forty years she happily baked my grandfather
rabbit pie, brawn, haslet;
collected fresh farm milk,
still-twitching pullets
and their warm muck-splattered eggs,
manure for the rhubarb, and mushrooms
dawn-gathered in chill Lincolnshire fields.

Judith Nicholls

Crying to get out

Inside every fat girl
There's a thin girl crying to get out
Sweet and sad and slinky
That nobody ever knows about

Inside every old man
There's a young man crying to get out
Just behind the wrinkles
Is the kid who used to twist and shout

It's the envelope that lies
Only look into their eyes
See the lonely dreamers there
Building castles in the air

Inside every hater
There's a lover dying of the drought
Inside every killer
There's a lover crying to get out
Trying to get out
Dying to get out
All the locked up lovers
Crying to get out

Fran Landesman

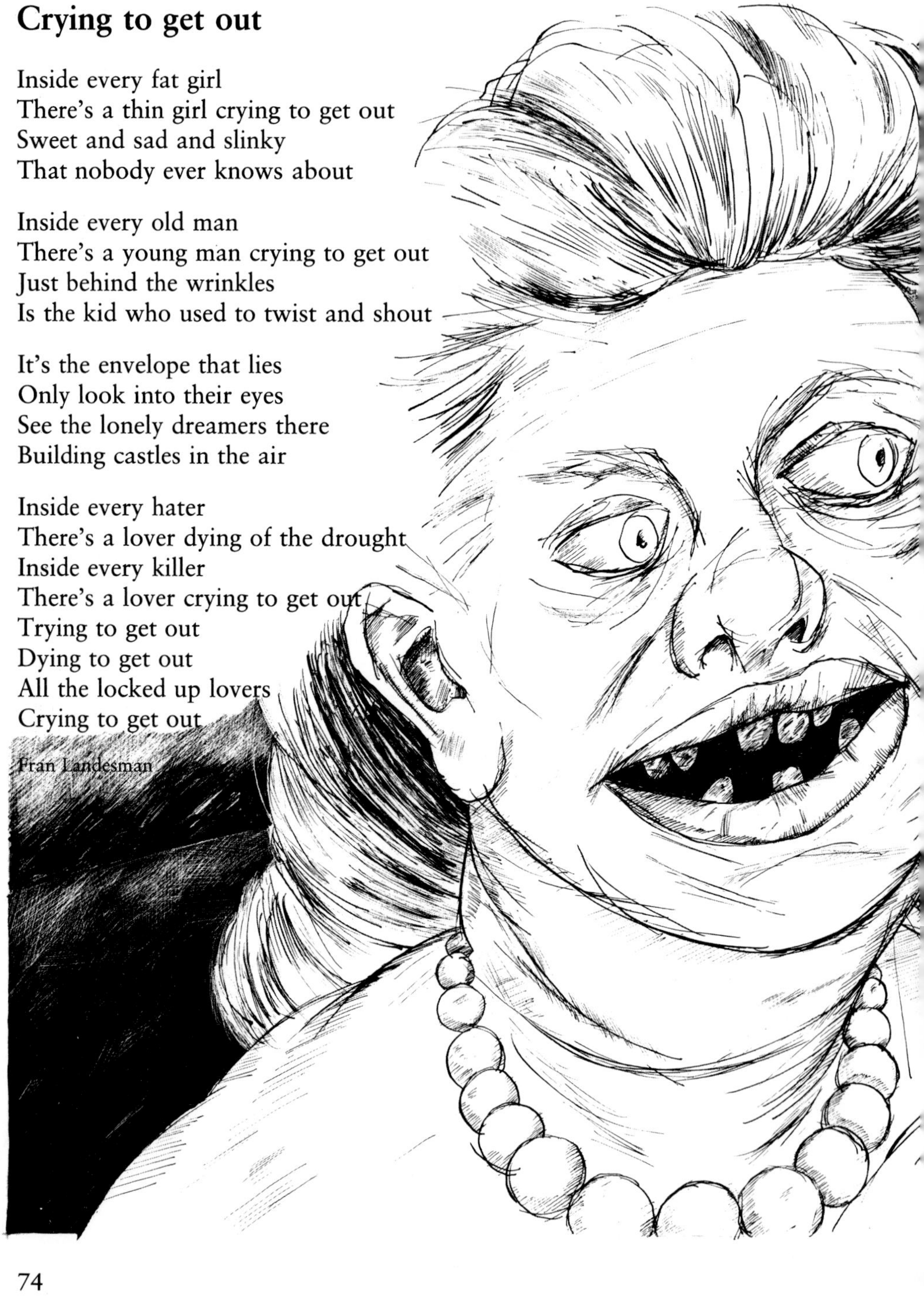

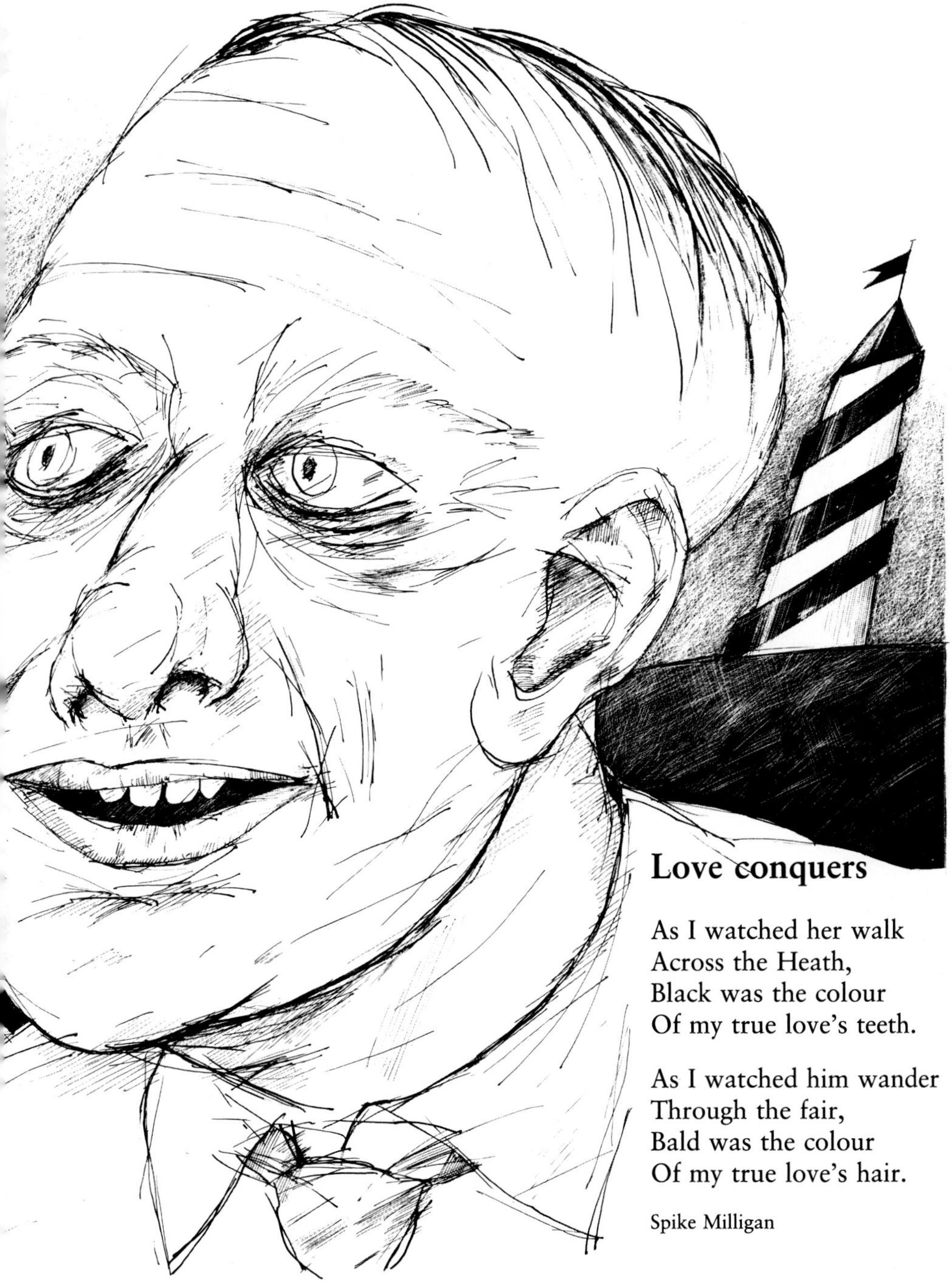

Love conquers

As I watched her walk
Across the Heath,
Black was the colour
Of my true love's teeth.

As I watched him wander
Through the fair,
Bald was the colour
Of my true love's hair.

Spike Milligan

On a country walk in a very thick mist

On a country walk in a very thick mist
My girlfriend asked if I'd give her a kiss,
  But I missed her lips
  Because of the fog
And I kissed her little one-eyed dog.

John Rice

Dreadful dream

I'm glad to say
I've never met
A cannibal
And yet

In this dream
I had last night
I almost took
A bite

Of somebody
Or other who
Was standing in
A stew –

Pot full of water
On a batch
Of firewood. Just
One match

Would do to start
It off. Whereat
I hollered, 'Just
Stop that!'

But cannibals
Began to shout:
'You've got to try
Him out!'

'Who? Me?' 'That's right.'
'I'd rather not.'
I said. They yelled,
'You've *got*

To have a taste!
Don't *want* to?' 'Not,'
I answered them,
'A lot,

Since, now you put
Me on the spot,
He wouldn't taste
So hot.'

At this the chap
Who'd soon be stewed
Remarked, 'I call
That *rude*!

Most ungrateful!
Here I am –
About to be
A ham –

Just to *please* you!
"WHAT A TREAT",
Not "HE'S UNFIT
TO EAT",

Is what I'd hoped
To hear. Well, stuff
It then, I've had
Enough.'

And out he stepped
As bold as day,
Shook-dried and stalked
Away.

The cannibals got
Quite excited:
'Last time *you're*
Invited!

Could have had him
On a bun:
Now look what
You've done!

Hurt his feelings,
BUNGLEHEAD!
We'll eat *you*
Instead.'

Things had got
Beyond a joke.
Thank goodness I
Awoke.

Dreadful dream!
Next time I sleep
I plan to do
It deep.

Kit Wright

My brother dreams of giants

My brother dreams the world is the size of a football
(It is kicked away by a giant.)
He dreams the sky is a large window
(The giant's child breaks it with a stone.)
He dreams he cannot hear his friends talking
(Beyond the clouds, the giant's dog is barking.)
He dreams there are no apples in any orchard in the world
(The giant's wife has picked them all.)
He dreams about the giant's family. He asks,
'Where do they live?'
'There are no giants,' says father.
'They've gone,' says mother.
My brother will not believe them.

He has seen the sky crack open.
He has heard the giant shouting.

'Where do the giants live!' he yells,
'Where do they live!'
'Is it beyond the sky? I know giants don't die!'

Bernard Logan

Creating light

A nine-year-old in bed
With magic – a bull's-eye torch,
His birthday present, – feels like
The Creator as he makes light
In the darkness. How easy:
Like saying 'Let there be Light!'
Indeed, searching the ceiling
With a focussed beam, he feels
He is creating more
Than light; chairs, shoes,
Shelves suddenly spring
From nothing into sharp focus.
'Let there be Light', he murmurs,
Knowing the first thing
Created, after heaven and earth,
Was light; but could that be?
The Mouth which gave that order
Created words first,
And so – 'Let there be Sound',
He mutters, half asleep ...

But then his imp sister
In the other bed wakes up
And shouts 'What's that you're doing?'
And the brother, scenting mischief,
Cries out, 'Look up, look up!
There's a spider on the ceiling;
Do you see? It's growing bigger.'

The wide-open hand
Over the open torch
Squirms and wriggles, casts
A frightening black shadow;
And as it looms closer
To the torch, the shadow swells,
Engulfs the whole room ...

The baby sister screams,
And the wicked brother chuckles;
'He'll eat us all up'.
And then to himself, darkly,
'I say "Let there be Darkness".'

Edward Lowbury

Time you did

Dawn is the only time
Birds get for rehearsing new songs.

Day is when people and vehicles
Cruise like lost submarines.

Evening could deafen you
If everyone turned their TVs up full volume.

Night is the only time
Shy animals get to window-shop

John Rice

Invasion

WITH THE FIRST
THE GULLS
FROM THE SEA
OVER
CAME BEATING IN
EDGE OF LIGHT

THE FARMLAND INTO
BIN COUNTRY. WAKING THE
TOWN
ROOFCOUNTRY, DUST-

FROM ITS SUNDAY MORNING BED THEY
FILLED THE AIR WITH THE
SCREAMS

FILLED THE PALE
OF THEIR DISSENSION,
SKY WITH THEIR ARROGANT
STRONG

WEAVING THEY BUILT
WINGS. WHEELING AND
A TOWERING PATTERN OF FLIGHT
ABOVE THE
TOWN.

Pamela Gillilan

The rooks

The bald-faced rooks,
Blown ragged,
Their fingered wings awry,
Harsh-voiced and vengeful
Claw the edge of the gale
And land in a jagged line.

Imperiously
They advance
Over the greening ground,
Dagger heads stabbing
The fragile shoots of wheat
Crouched to earth in the wind.

A shot vibrates.
Abruptly
Their wings blacken the air.
One lifeless comrade
Is left as they cluster
Funereal in the far-off trees.

Albert Rowe

Wood pigeon

Wood pigeon
makes a black arrow on the cloud
and a procession between the wheat,
beats the leaves
of the trees it leaves,
crosses the sun
and gargles in a hedge.

The pigeons in town
race in laps round the ABC cinema
swell on swiss-rolls
and go about in gangs
robbing sparrows.

Joe one struts down our gutter
every morning
and just beneath the window –
gargles.

Anyone would think it was a hedge.

Michael Rosen

Insecure trouser-nesting hairee

You've probably never heard
Of the trouser-nesting bird.
He lives on tops of houses
And wears his nest like trousers.

Spike Milligan

The hedgehog has Itchy the hedgehog to hug

The hedgehog has Itchy the hedgehog to hug
And a hedgehog bug has a hedgehog bug.

Hedgehog with hedgehog is happy at ease
And hedgehogs with fleas, and fleas with fleas.

The batch of the flea's eggs hatch in the crutch
Of the hedgehog's armpit, a hot rich hutch.

The hedgehog's clutch of hoglets come
In the niche of a ditch, from the hedgepig's tum.

And so they enjoy their mutual joke
With a pricklety itch and a scratchety poke.

Ted Hughes

The new gnus

A gnu who was new to the zoo
Asked another gnu what he should do.
The other gnu said, shaking his head,
'If I knew, I'd tell you. I'm new too!'

John Foster

How to tell a camel

The **D**romedary has one hump,

The **B**actrian has two.

It's easy to forget this rule,

So here is what to do.

Roll the first initial over

On its flat behind:

The **B**actrian is different from

The **D**romedary kind.

J. Patrick Lewis

The defence

A silent murderer,
A kestrel came today,
Canny marauder
In search of prey.

He scanned the ground for signs,
Hovered on the lifting air,
His claws poised ready to snatch
Some victim off to his lair.

But swallows gathered fast,
Closed in from all around,
Sounding a shrill alarm,
Filling the sky with furious sound.

They soared into the sky,
And peeled off to attack,
Jabbing with dagger beak and needling cry
To drive the killer back.

The kestrel wavered once,
Then, like a useless glove
Thrown down, he tumbled on the wind
And fled from the birds above.

And we who sat and watched,
We too had been afeared,
Had held our bated breath.
But now we stood up and cheered!

Geoffrey Summerfield

Visitor

Sliding in slippers along the house-side
You find fragments of the turkey's carcass
Beside the toppled dustbin lid, and there
On the lawn's snow quilt, a line of paw marks.

Town fox, that wraith of winter, soundlessly
Thieved here as frost bit hard and stars shivered.
This bleak morning, under a raw-boned sky,
You stoop to examine the frozen tracks.

And print yours where a spectral guest came late
To share a Christmas dinner. Around the
Gable-end a starved wind razors and from
The split gutters icicles hang like fangs.

Wes Magee

Sheep

Stuck in a field, grazing,
The sheep seem quite content.
They look tired from all their lazing,
Their heads are always bent.

They know their place, they seem
To perform a simple role;
They crop grass as a team,
They are easy to control.

Or are they? Who says so?
Are they taken for granted?
They may be eager to show
They are far from contented.

Imagine changing places with them,
Taking their place;
Imagine being a little lamb,
Think of living on grass.

Imagine hearing what they'd say
If they could only speak:
'We've had a baaad time today,
Give us a break.'

If they could lord it over us,
If they took us by surprise,
I doubt if they'd bother us,
Pull the wool over our eyes.

I doubt if they'd think
We were good enough to eat;
I doubt if they'd see us
As so much butcher meat.

I doubt if they'd sell us
For cash or for cheques;
I can't see them taking
The coats from our backs.

I think they'd ban cars
   And bring all fences down;
They'd want none of our wars,
   They'd keep clear of each town.

I think they'd just wander
   And leave us behind.
They'd forget all our plunder:
   Out of sight, out of mind.

Alan Bold

Rabbits

The last spinney of wheat,
The clacking reaper making its turn,
And the guns ready.
We sat close to the hedge
By the pile of jackets and half-empty
Bottles of cold tea, breathless
With fear for the creatures still crouching
Among the thinning stalks, and then
Half-lusting for their deaths.
Broken out they made their rushing bid
For the safe ditch, died leaping
As if struck by a fist of air.
Like the victims of a disaster
They were laid in rows
Limp and bloody-nosed, all speed
And strength gone out of them.

Pamela Gillilan

The sett

Where the crowded rows of conifers
Link their lower leafless arms one with another
And sunlight on fallen branches shines as rare
As on sunken treasure, where wood pigeons panic
At a snapping twig and the path once cleared
Is blocked by trees pulled down by the storm,
The shy badger tunnels deep down
Away from the faintest light or sound
To the perfect dark and quiet underground.

Stanley Cook

Dream of the fair forest

Where can I hide in you,
forest, fair forest?
Where can I hide in you,
forest so green?

Hide in my pine trees,
my beeches, my aspen.
Hide in my maples,
where none can be seen.

How shall I live in you,
forest, fair forest?
How shall I live in you,
forest so gay?

Live on my berries
my cobnuts, my rosehips.
Live on my blackberries,
They'll last many-a-day.

Who shall I live with,
oh forest, fair forest?
Who shall I live with,
my forest so wild?

Live with my squirrels,
my nuthatch, my night owls.
Live with my badgers
and live as my child.

What shall I lie on,
green forest, my forest?
What shall I lie on
when night starts to fall?

Lie on my grasses,
my rosebay, my lichen.
Lie on my mosses
the softest of all.

What if I'm lonely,
fine forest, fair forest?
What if I dream
of returning to town?

You won't be lonely,
fair dreamchild, my wanderer.
With fox cub and grey dove
you won't be alone.

Judith Nicholls

Goliath

They chop down 100ft trees
To make chairs
I bought one
I am six-foot one inch.
When I sit in the chair
I'm four-foot two.
Did they really chop down a 100ft tree
To make me look shorter?

Spike Milligan

The scarecrow

The scarecrow is a scarey crow
Who guards a private patch
Waiting for a trespassing
Little girl to snatch

Spitting soil into her mouth
His twiggy fingers scratch
Pulls her down on to the ground
As circling birdies watch

Drags her to his hidey-hole
And opens up the hatch
Throws her to the crawlies
Then double locks the latch

The scarecrow is a scarey crow
Always out to catch
Juicy bits of compost
to feed his cabbage patch

So don't go where the scarecrows are
Don't go there, Don't go there
Don't go where the scarecrows are
Don't go, Don't go ...

Don't go where the scarecrows are
Don't go there, Don't go there
Don't go where the scarecrows are
Don't go ...

Roger McGough

Haunted

Black hill
black hall
all still
owl's grey cry
edges shrill
castle night.

Woken eye
round in fright;
what lurks walks
in castle rustle?

Hand cold
held hand
the moving roving
urging thing:
dreamed margin

voiceless
noiseless
HEARD
feared
a ghost passed

black hill
black hall
all still
owl's grey cry
edges shrill
castle night.

William Mayne

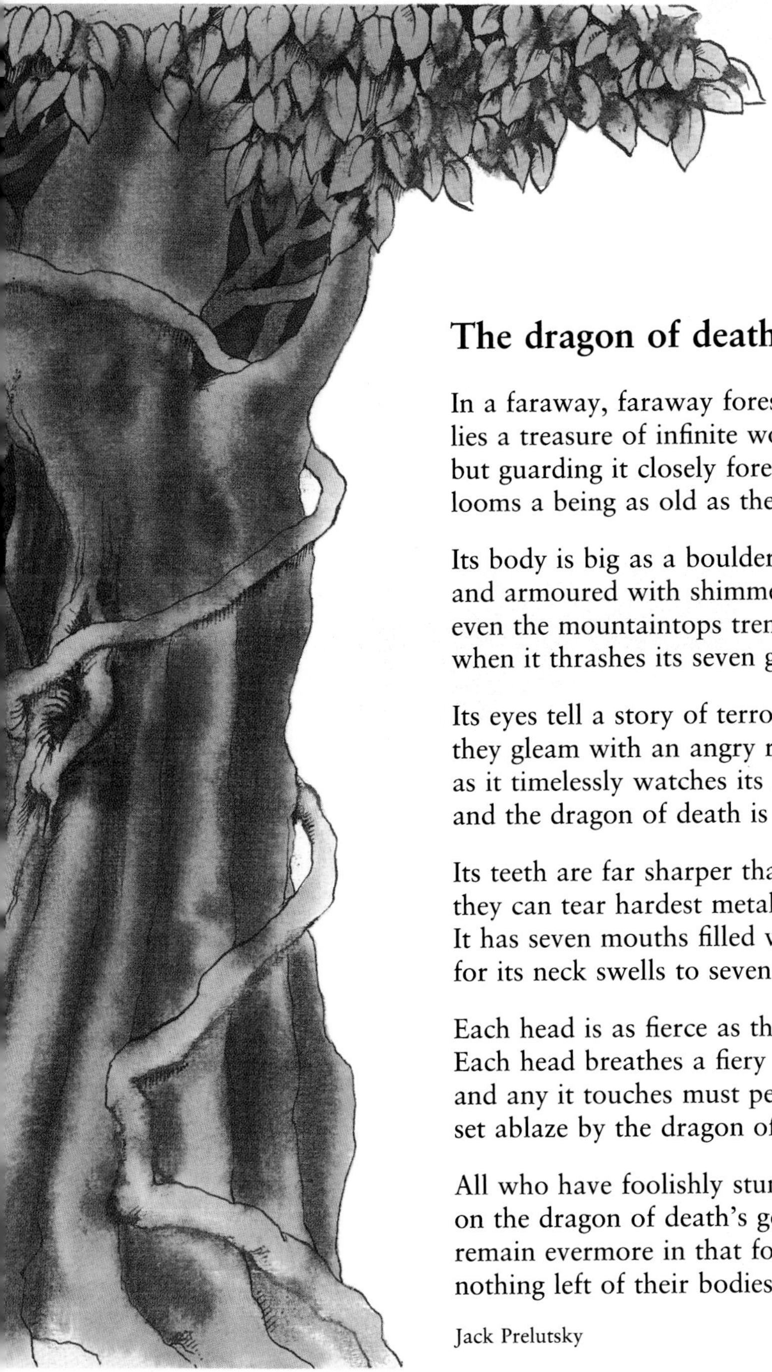

The dragon of death

In a faraway, faraway forest
lies a treasure of infinite worth,
but guarding it closely forever
looms a being as old as the earth.

Its body is big as a boulder
and armoured with shimmering scales,
even the mountaintops tremble
when it thrashes its seven great tails.

Its eyes tell a story of terror,
they gleam with an angry red flame
as it timelessly watches its riches,
and the dragon of death is its name.

Its teeth are far sharper than daggers,
they can tear hardest metal to shreds.
It has seven mouths filled with these weapons,
for its neck swells to seven great heads.

Each head is as fierce as the other,
Each head breathes a fiery breath,
and any it touches must perish,
set ablaze by the dragon of death.

All who have foolishly stumbled
on the dragon of death's golden cache
remain evermore in that forest,
nothing left of their bodies but ash.

Jack Prelutsky

Away from it all

Conjure up a whole new moon,
A landscape loud with hungry owls,
A road that climbs where mountains soar
And separate from hills.

October: and a car climbs up
The mountain, staggers on the slope,
Gives up, oh roughly halfway there.
The snoring engine goes to sleep.

A family clambers out: a man,
A woman and a boy and girl;
They're lost without their zonked-out car.
They feel the wind creep up and snarl.

They look around: there's nothing there,
Well, nothing that they recognise;
They hear the thunder, then they feel
The downpour falling from the skies.

They seek a shelter, trudge along the road,
They want a light to match the beaming moon;
They march in silence for a mile,
Though they are four each one's alone.

Another mile, and then they see
A beacon in the distant dark;
Their footsteps quicken, they approach
A mansion in a massive park.

Although it's lit, the place looks odd:
The door is open; no one's there.
They enter, shout a greeting, get
An echo from the empty air.

'Ah well,' says father, 'it'll do.
  We need a shelter for the night.'
They take two rooms, the beds are made;
  They fall asleep in seconds flat.

Under the moon the mansion lifts,
  It hovers in the gloom awhile;
Bright lights blaze out, a sound escapes,
  The whisper of a whirring wheel.

Then up the structure goes, straight up
  And launches through the atmosphere.
They'll wake up where they started out,
  A zillion millions from anywhere.

Alan Bold

The game ... at the Hallowe'en party in Hangman's Wood

Around the trees ran witches
  Their nails as long as knives.
Behind a bush hid demons
  In fear for their lives.

Murder, murder in the dark!
The screams ring in your ears.
It's just a game, a silly lark,
No need for floods of tears.

Tall ghosts and other nasties
  Jumped out and wailed like trains.
A skeleton in irons
  Kept rattling his chains.

Murder, murder in the dark!
The screams ring in your ears.
It's just a game, a silly lark,
So wipe away those tears.

A werewolf howled his heart out;
  The Horrid Dwarf crept by.
There was blood upon his boots
  And murder in his eye.

Murder, murder in the dark!
The screams ring in your ears.
It's just a game, a silly lark;
Oh, come now, no more tears.

Owls were hooting, 'Is it you?'
  Until a wizard grim
Pointed to the Dwarf and said,
  'The murderer ... it's him!'

Murder, murder in the dark!
The screams ring in your ears.
It's just a game, a silly lark;
There's no time left for tears.

Murder, murder in the dark!
The screams fade in the night.
Listen, there's a farm dog's bark!
And look, the dawn's first light!

Wes Magee

Empty fears

What's that? – Coming after me, down the street,
With the sound of somebody dragging one foot
Behind him, who pauses, who watches, who goes
With a shuffle and mutter
From the wall to the gutter
In the patch where the light from the lamps doesn't meet...

Oh ... it's only a bit of paper – a hollow brown bag
Open-mouthed, like a shout – a bit like the face
Crumpled-up, of someone who's going to cry,
Blown on the wind, from place to place,
Pointless, and light, and dry.

Who's that? – Watching, from the upstairs windows
Of the house where the hedge grows right back to the door,
Where the half-drawn curtains droop and discolour
And a yellow bulb burns away
And the milk's on the step all day –
Somebody lives there, no one comes or goes ...

Oh ... it's only an empty coat on a hanger
That sways in a draught like a man who depends
On only one thing – the something inside
That's holding him up, waiting for friends
He writes to, but no one's replied.

What's that? – Whispering, where the fence round the lot
Sags like a fading hope: the gate just here twists
On its hinge like a bird's broken wing
And shrieks as you look, and see:
Nothing, where all the shops used to be,
People coming and going where now they are not...

Oh ... it's only the breeze, that's fretting itself
Amongst the stiff thistles, each standing alone,
Upright, all winter, dead, but not gone...

But if it's only these things, what blows
Through me, to make me afraid, who knows?

Brian Lee

Footsteps

Footsteps echo
Down empty halls
Drumming a rhythm
  That mocks
  At locks
  And cocks
  A snook
At rational thought.
  Ought
  There
Not be a
Pair of feet
Causing that beat?
Or is the echo
  I hear
  My fear?

Vicky Blake

Canal lock in winter

They stood by the bank and called me names;
'Yaller', they screamed and laughed like knives,
Pulled at their socks and blew their cheeks,
And pretended to split the ice with dives.

They thumbed their noses and gloated in dance,
And wagged their fore-nails with a sneer,
And guarded the gates and the way across –
'Fatty's scared to get too near'.

I'd watched them feather from wall to wall,
Powder tight-toed through creaking snow,
And I was last –
And lost in fear
Of the twelve-foot blackness of water below;

And the green-slimed cliffs and the frozen hiss
Of the water whiskering from the gate,
And the dozen watchful, mocking eyes
Grinning with hate.

With all my courage left I ran,
Scrambled the bank and rolled the wire,
And from the lock their anger rose
And scorched my running like a fire.

Gregory Harrison

Child with toy sword

Clutching in hero hand the bright
Symbol of death and glory,
He marches bravely to the fight
And grows as tall as a story.

The garden birds blow bugles for
His joy as he advances;
The sun declares a playful war
And throws down flashing lances.

And no one whispers to the boy
That some hot, future afternoon
He will lunge upward with that toy
And burst the sun like a huge balloon.

Vernon Scannell

Toy wars

Whether it rains
or shines, you play
the livelong day
with trucks and trains
with cars and cranes.

From door to door
and floor to floor
your cannons roar
your bombers soar
your robots fire flames
in never-ending video games
of hi-tech computer wars.

I'm just a bit afraid
that one fine day
they'll start to play
without your aid
the wars you've made.

From door to door
and floor to floor
your cannons roar
your bombers soar
your robots fire flames
in never-ending video games
of hi-tech computer wars.

While you, a helpless girl or boy
lie in a corner like a broken toy.

James Kirkup

TV news

I sat and ate my tea
And watched them die
And knew at twelve
That brave men died
On either side.
And that is why
At twelve I cried – and cried,
And couldn't understand
My father's beaming pride.

John Kitching

If I die in War

If I die in War
You remember me
If I live in Peace
You don't.

Spike Milligan

Lost in the museum

Every glass case in a museum
Is an island surrounded by time.
To reach them you have to imagine
Crossing not seas but centuries
To a different kind of life.

Separated from us not by waves
But all the years that have passed
Are places where they still run
The first steam locomotive
And cheer the speedmen on
In penny farthing bicycle races.

And there with the sickle to cut the corn
And millstones to grind it to flour
Is the kind of shelter we'd build
Were we lost on a distant island
And those are the tools we'd invent
If no one ever came to find us.

Stanley Cook

The lion makers

There were four youths in India
Two thousand years ago
Whose story's the truest and most up to date
Of any story I know.

Three of the four were smart as paint,
Enough to make you sick,
With top marks in everything they did,
Especially arithmetic.

Just one of the four (Gumption by name)
Was more like the rest of us:
Average in most things, practical,
Far from a genius.

On leaving school, 'What shall we do?'
The clever trio cried.
'Why, make your fortune by your brains,'
Gumption, their friend, replied.

'But how?' they asked, 'And where? And when?'
So nothing at all got planned.
Gumption, deciding for them, led
Them off to a distant land.

For an Emperor good and wise ruled there,
Whose greatest personal pleasure
Was in seeking out men of genius
And loading them with treasure.

On their way they passed through a jungle glade
Where bones lay bleached in the sun.
The first of the trio halted and gathered
Them thoughtfully, one by one.

He assembled skull, kneecap and spine,
Jawbone and shoulder-blade,
Till the skeleton of a lion lay
Spread out in the jungle glade.

His eyes shone with challenge and triumph,
So the second of the three
Sloped off to the jungle, and soon came back
With his brows bunched thoughtfully.

In his arms he carried a dead weight
Of cartilage, hair, gut, hide,
Which he bound and wound round the skeleton
And secured and deftly tied.

And his eyes shone with challenge and triumph,
For under the glare of the day,
Stretched out, and perfect in all of its parts,
A lifeless lion lay.

So the third of the trio stood with bunched brows,
Chewing his thumb thoughtfully,
Then a glint of triumph shone in his eyes
And he lowered himself on one knee.

He took in his hands the great slumped head –
At which point, Gumption flew
And climbed the nearest tree to watch
What fresh marvel might ensue.

Not to be outdone, the third of the three
Through its nostrils, dry and pale,
Breathed a secret code that shuddered the beast
From mane to lashing tail.

It leaped to life, it killed all three,
It ate them one by one,
And left their bones in the jungle glade
To bleach in the glare of the sun.

So Gumption lived to tell the tale;
He scrambled down from the tree
And ever after was careful to keep
Out of bad company.

This story's all too up to date,
Nor is it hard to know
The lion-makers of today,
Whose numbers grow and grow.

What's different is that the jungle's gone,
So Gumption's in a worse plight:
There are lions prowling flat skylines,
But never a tree in sight.

Raymond Wilson

The lost echo

At the edge of the precipice
a weather-beaten wooden sign
whispers in faded characters:
'This was the site of
the celebrated multiple echo
that once attracted to our land
thousands of tourists and honeymooners.'

Every day they would come
to picnic and throw their voices
at the mysterious answerer
in a bedlam of tongues.
Whatever language they spoke,
he was fluent in it, and always
replied with a flawless accent.

Questions, prayers, handclaps, laughter,
top C's, nicknames, slogans, inane remarks –
in favourable conditions, the echo's mimic
had them all by heart, as much as seven times,
clear as a bell – a bell a bell a bell:
even ten times was not unknown –
not unknown – not unknown unknown.

But some fifty years ago – to this day
no one has been able to figure why –
the echoes began to deteriorate, decay:
the responses became muted, faded,
and their numbers, like the visitors,
declined from as many as ten
to five, three, one – then none.

For even that single
confused echo slowly died away into itself,
and was never heard again.
Even the greatest coloratura sopranos
were unable to resuscitate it,
and the Imperial Band of the National Guard
had to beat a retreat.

The answerer among the heights had departed,
with all his multiple voices, leaving
this uninhabitable void –
Only one old man still comes here every day,
a last survivor, to throw his weakening voice
against the void –
but throws his voice in vain.

Where has our echo gone?
How did its Tower of Babel fall?
– Some say the voice lies buried
deep in the heart of a mountain lake,
unknown to man – unknown to man –
and cannot be resurrected till the day
the day when silence is the universal tongue.

Then the mysterious answerer
will rise up from the lake
with re-animated heart and mind, and –
listening to the primal stillness
to the stillness in men's souls,
and to the peace within their hearts –
will draw a profound breath

will draw a profound breath
and set the mountains ringing,
and all our devastated land,
and set the whole world singing,
and all the universes, all
the universes, all, with silent echoes –
echoes of stillness – stillness stillness stillness –

James Kirkup

Tramp

I could see he was a bearded brigand
With a fierce, full spade
Of whiskers,
Bush-ranger hat
And sleeveless, leather jacket.

The dry twigs underneath the hedge
Cracked, writhed and twisted
In the red-hot wigwam of his healing fire.
Underpants and vest
Hung on a thin, stretched line
Over the halberds of the yellow flames,
Trousers softening to dryness on his hairy legs,
Feet simmering in the sweat of steaming socks.

It will rain,
The bundles of the black, low clouds make sure.
But in the spinney is a simple hut,
An oilskin roof,
Some plastic bags,
Dry sacks,
And purring on the fire a pan
For cocoa
And a stone bed-bottle.

But all too soon,
Hours before yellow dawn lights up the fret
Of branches in the hedge,
His knees will lock with cold;
The fire will settle into grey, wet ash.
A spinney is a harsh, bleak place
To lie
All night in rain.

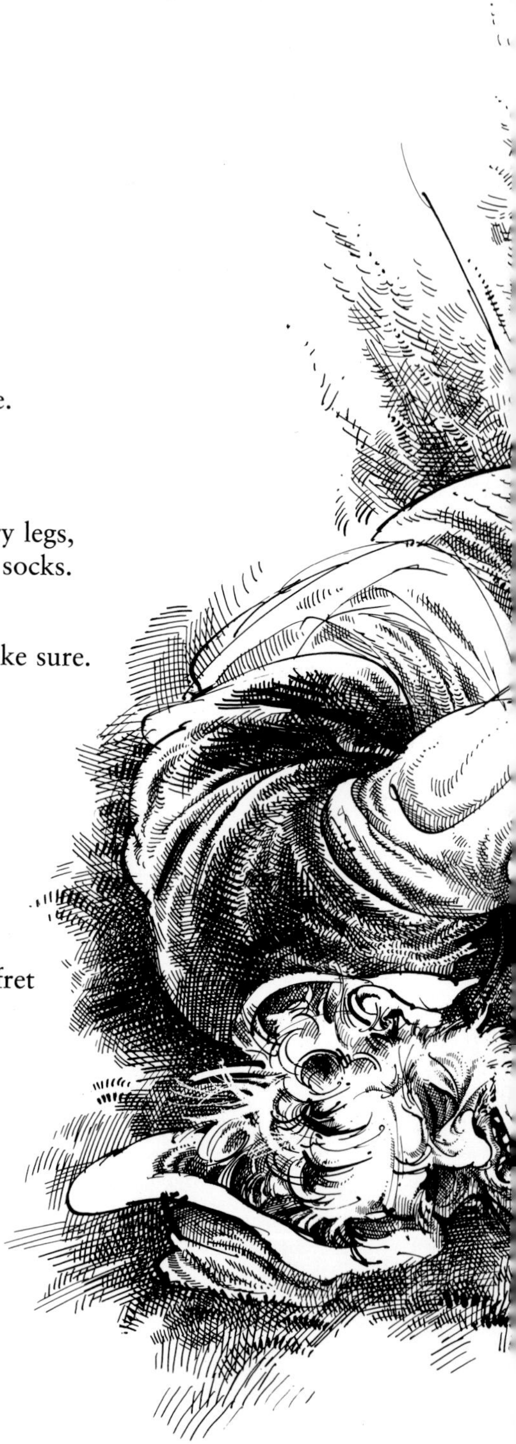

And in the morning
He will stagger through stiffness,
Lean on the handles of the grubby pram
Stuffed with his clothes, his bedding
And his clanking pans.
His medals tinkle on their faded silks,
And sometimes he will talk
Of this campaign or that –
Nothing heroic,
He was there, that's all.
And by his accent and his ready words
There are romantics who would judge him
Class,
College at least perhaps, or university –
A sort of scholar gipsy.

One day,
Savaged by freedom
Turned to a snarling master,
They found him by the tyrant,
Stretched in a ditch –
One afternoon
A farm-hand and a policeman on patrol
Stood by the shrunken foetus,
Waxen, already cold,
Beard still defiant in sleep.

Gregory Harrison

Her garden

She was too old to stop weeds trespassing
On her garden, whether they parachuted in
Like the willow herb and dandelion seeds
Or like the brambles and goosegrass,
Roped in and using thorns and hooks
For climbing irons, came over the wall.
She and her cat as big as a dog
Hadn't the energy to stop
The daisies and buttercups
Making themselves at home on the lawn
And foxgloves in their steeple hats
Behaving as if they owned the place.

She looked at us first from behind the curtain
When we called one afternoon
To do our bit of tidying up
And watched us trip the weeds with a hoe
And snap at nettles with the shears.
All kinds of wild flowers ran wild
And sycamore seeds storms had flown in
Spinning like rotors of helicopters
Were turning into tiny trees
And, given time, the house would have been
Surrounded by a forest.

We threw weeds out and scared them off
Till you saw the roses that before
You couldn't see. The garden was really nice
And so was she.

Stanley Cook

The doll

'She is older than I am',
Chuckles the old woman,
Pointing a shrivelled finger
At the child propped in a pram
Bolt-upright, blinking
When budged, its chubby face
Smooth and expressionless –
The doll she had for Christmas
Ninety years ago.
Unearthed from a dark cupboard,
A toy which was too big
For her, but did not grow,
Now comes into its own.
'It was so tall', she muses,
'So much like me, people
Mistook it for my twin'.
And now, bridging the gap
Of ninety years, she sees
A twin's eyes look back;
Slowly forgets the map
Of rivulets and wrinkles
Through which she talks to us –
And the doll seems to smile
Back from a looking-glass.

Edward Lowbury

The future

The young boy stood looking up the road
to the future. In the distance both sides
appeared to converge together. 'That
is due to perspective, when you reach
there the road is as wide as it is here',
said an old wise man. The young
boy set off on the road, but,
as he went on, both sides of the
road converged until he could
go no further. He returned to ask
the old man what to do, but
the old man was dead.

Spike Milligan

Walls

Man is
a great wall builder
the Berlin Wall
the Wailing Wall of Jerusalem
but the wall
most impregnable
has a moat
flowing with fright
around his heart.

A wall
without windows
for the spirit
to breeze through

A wall
without a door
for love to walk in.

Mbuyiseni Oswald Mtshali

Song of hope

At that hour
when the sun
slinks off
behind hills
and night
– a panther –
crouches
ready to spring
upon our un-
suspecting city . . .

i want to sing
the coiled desires
of this land
the caged dreams
of forgotten men

i want to sing
of all that was
but no longer is
of all that
never was but
could have been

i want to sing
the obsidian
unspelled hopes
of our children
i want to sing
to remind us
never to despair
that every hour
every minute
somewhere on the face
of this earth
it is glorious morning.

Cecil Rajendra

Index of first lines

Acknowledgements

The following poems are appearing for the first time in this collection and are reprinted by permission of the author unless otherwise stated.

Les Arnold: 'The sum' © 1985 Les Arnold. Leo Aylen: 'Stage struck' © 1985 Leo Aylen. Alan Bold: 'And now'; 'Sheep' and 'Away from it all'. All © 1985 Alan Bold. Keith Bosley: 'Guardian' © 1985 Keith Bosley. Stanley Cook: 'Racing cyclist'; 'The sett'; 'Lost in the museum' and 'Her garden'. All © 1985 Stanley Cook. John Cunliffe: 'Grown up' © 1985 John Cunliffe. Eric Finney: 'Conversation with insults'; 'Sarky devil'; 'Poem about writing a poem' and 'Billy and me and the pane'. All © 1985 Eric Finney. John Foster: 'Four o'clock Friday'; 'There are four chairs round the table'; 'Aunty Joan'; 'Great-Gran' and 'The new gnus'. All © 1985 John Foster. Pamela Gillilan: 'July evening'; 'Haiku calendar'; 'Poetry (or cricket)'; 'Mottled moths'; 'Rabbits'; 'The half-mile' and 'Invasion'. All © 1985 Pamela Gillilan. David Harmer: 'Next door' and 'On a blue day'. Both © 1985 David Harmer. Gregory Harrison: 'The cave'; 'The cornfield den' and 'Tramp'. All © 1985 Gregory Harrison. Ted Hughes: 'The hedgehog has Itchy the hedgehog to hug' © 1985 Ted Hughes. James Kirkup: 'Summer full moon'; 'Toy wars' and 'The lost echo'. All © 1985 James Kirkup. John Kitching: 'Books'; 'One parent family'; 'Living on' and 'TV news'. All © 1985 John Kitching. Brian Lee: 'We say' and 'Empty fears'. Both © 1985 Brian Lee. J. Patrick Lewis: 'How to tell a camel' © 1985 J. Patrick Lewis. Edward Lowbury: 'Doctor Christmas'; 'Santa'; 'Creating light' and 'The doll'. All © 1985 Edward Lowbury. Wes Magee: 'Xmas'; 'School field in winter'; 'Still'; 'Visitor' and 'The game . . . at the Hallowe'en party in Hangman's Wood'. All © 1985 Wes Magee. Judith Nicholls: 'Starfish'; 'Ars mathematica'; 'Haiku for autumn'; 'West wind'; 'Uncle William'; 'Then' and 'Dream of the fair forest'. All © 1985 Judith Nicholls. Rony Robinson: 'Who is you? © 1985 Rony Robinson. Albert Rowe: 'The rooks' © 1985 Albert Rowe. John Rice: 'on the doorstep'; 'On a country walk in a very thick mist'; 'A chemistry student from Gillingham'; 'Muntze of the ear' and 'Time you did'. All © 1985 John Rice. Adrian Rumble: 'Autumn' © 1985 Adrian Rumble. Vernon Scannell: 'My Uncle Justin' and 'Child with toy sword'. Both © 1985 Vernon Scannell. Raymond Wilson: 'The lion makers' © 1985 Raymond Wilson.

The editor and publisher are grateful for permission to reprint the following poems:

Allan Ahlberg: 'Glenis' from *Please Mrs. Butler* (Kestrel Books 1983) p. 23. Copyright © 1983 by Allan Ahlberg. Reprinted by permission of Penguin Books Ltd. Joan Batchelor: 'Grandma' from *Home Truths*. Reprinted by permission of the author. Vicky Blake: 'Footsteps' from *Through Small Windows*. Charles Causley: 'Uncle Stan' from *Secret Destinations* (Macmillan, 1984), first published in Poetry Book Society Supplement 1981. Copyright © Charles Causley 1981. Reprinted by permission of David Higham Associates Ltd. Robert Froman: 'Quiet secret' from *Seeing Things*. Reprinted by permission of the author. Mick Gowar: 'Pocket money' © 1985 Mick Gowar. Reprinted by permission of William Collins Sons & Co., Ltd. Gregory Harrison: 'Canal lock in winter' from *Posting Letters* (OUP). Reprinted by permission of the author. Ted Hughes: 'April birthday' from *Season Songs*. Copyright © 1973, 1975 by Ted Hughes. Reprinted by permission of Faber & Faber Ltd., and Viking Penguin, Inc. Maxine Kelly-Robinson: 'Grand-parents' from *Hey Mister Butterfly*, ed. Alasdair Aston, p. 97 (ILEA, English Centre). Fran Landesman: 'Crying to get out' © Fran Landesman 1981, from *Golden Handshake*. Reprinted by permission of Jay Landesman Ltd. Laurie Lee: 'Christmas Landscape' from *A Bloom of Candles*. Reprinted by permission of the author. Philip Larkin: 'The trees' from *High Windows*. Copyright © 1974 by Philip Larkin. Reprinted by permission of Faber and Faber Ltd., and Farrar, Straus & Giroux, Inc. Bernard Logan: 'My brother dreams of giants' from *Gangsters, Ghosts and Dragonflies*, ed. Brian Patten. Reprinted by permission of Allen & Unwin (Publishers) Ltd. Edward Lowbury 'Unsecret' from *Green Magic* (Chatto & Windus). Reprinted by permission of the author. William Mayne: 'Haunted' from *Ghosts* (Hamish Hamilton). Reprinted by permission of David Higham Associates Ltd. Roger McGough: 'The scarecrow' from *New Volume*, ed. Brian Patten, Roger McGough and Adrian Henri (Penguin Books Ltd). Reprinted by permission of A.D. Peters & Co. Ltd. Spike Milligan: 'The future'; 'If I die In War' and 'Goliath' from *Small Dreams of a Scorpion* (Michael Joseph/M & J Hobbs); 'Love conquers' from *Unspun Socks from a Chicken's Laundry* (Michael Joseph/M & J Hobbs) and 'Insecure trouser-nesting hairee' from *A Book of Bits or A Bit of a Book* (Dobson). All reprinted by permission of Spike Milligan Productions Ltd. Mbuyiseni Oswald Mtshali: 'Sunset' and 'Walls' from *Sounds of a Cowhide Drum* © Mbuyiseni Oswald Mtshali 1971. Reprinted by permission of Oxford University Press. Jack Prelutsky: 'The dragon of death' from *Nightmares: Poems to Trouble Your Sleep*. Copyright © 1976 by Jack Prelutsky. Reprinted by permission of A & C Black Ltd., and Greenwillow Books (A Division of William Morrow). Cecil Rajendra: 'Leave-taking' and 'Song of hope' from *Hours of Assassins and Other Poems*. Reprinted by permission of Bogle-l'Ouverture Publications Ltd. Michael Rosen: 'Wood pigeon'. Reprinted by permission of the author. Geoffrey Summerfield: 'The calendar of my year' and 'The defence' from *Welcome and Other Poems*. Reprinted by permission of Andre Deutsch Ltd. Martyn Wiley: 'Cycle song' from *Versewagon*. Reprinted by permission of the author. Kit Wright: 'Dreadful dream' from *Hot Dog and Other Poems* (Kestrel Books 1981) pp. 27–31. Text copyright © 1981 Kit Wright. Reprinted by permission of Penguin Books Ltd.

Illustrations are by Peter Benton, Allan Curless, Tudor Humphries, Robert Kettell, David Parkins, Kate Simpson.

Cover illustration is by Allan Curless

The publisher would like to thank the following for permission to reproduce photographs:

Bodleian Library p.52, p.53; J. Allan Cash p.54, p.55, p.59, p.84, p.93; Jane Davies p.124; Natural History Photographic Agency: Stephen Dalton p.45, Roy Shaw p.57; Reflex: Ashdown/Gordon p.107, Cavendish p.110, p.111; Adrian Smith p.9.